Name _____ Class _____ Date _____

Skills Worksheet

Directed Reading

Section: Chromosomes

In the space provided, explain how the terms in each pair differ in meaning.

1. cell division, gamete

_____

_____

_____

_____

2. gene, DNA

_____

_____

_____

_____

3. chromosomes, chromatids

_____

_____

_____

_____

Study the following steps of binary fission in a bacterium. Determine the order in which the steps take place. Write the number of each step in the space provided.

_____ **4.** New cell wall forms around the new membrane.

_____ **5.** New cell membrane is added to a point on the membrane between the two DNA copies.

_____ **6.** The bacterium is pinched into two independent cells.

_____ **7.** The growing cell membrane pushes inward, and the cell is constricted in two.

_____ **8.** DNA is copied.

Complete each statement by writing the correct term or phrase in the space provided.

9. Chromosomes that are similar in size, shape, and genetic content are called

    _____ .

10. A cell, such as a somatic cell, that contains two sets of chromosomes is said

    to be _____ .

11. Biologists use the symbol _____ to represent one set of
    chromosomes.

12. A fertilized egg cell, the first cell of a new individual, is called a(n)

    _____ .

Read each question, and write your answer in the space provided.

13. What is the difference between an autosome and a sex chromosome?

    _____

    _____

    _____

14. What is a karyotype?

    _____

    _____

    _____

15. Describe four types of mutations resulting from the breakage of chromosomes.

    _____

    _____

    _____

    _____

Name _____ Class _____ Date _____

Directed Reading

Section: The Cell Cycle

Complete each statement by writing the correct term or phrase in the space provided.

1. The cell cycle is a repeating sequence of growth and

   _____ during the life of a cell.

2. The first three phases of the cell cycle are collectively called

   _____ .

In the space provided, write the letter of the description that best matches the term or phrase.

_____ 3. first growth (G_1) phase

_____ 4. synthesis (S) phase

_____ 5. second growth (G_2) phase

_____ 6. mitosis

_____ 7. cytokinesis

a. nucleus divides into two nuclei

b. cytoplasm divides

c. preparations are made for the nucleus to divide

d. DNA is copied

e. cell carries out its routine functions

Complete each statement by writing the correct term or phrase in the space provided.

8. Many _____ control the cell cycle.

9. The checkpoint that makes the key decision of whether the cell will divide is

   the _____ _____ checkpoint.

10. The information necessary to make the proteins that regulate cell growth and

    division is contained in _____ .

Read the question, and write your answer in the space provided.

11. Describe the role of checkpoints in the onset of cancer.

   _____

   _____

   _____

Skills Worksheet

Directed Reading

Section: Mitosis and Cytokinesis

Read each question, and write your answer in the space provided.

1. What function do spindles perform during mitosis?

_____

2. What function do centrioles perform in animal cell mitosis?

_____

In the space provided, write the letter of the description that best matches the term or phrase.

_____ **3.** prophase

_____ **4.** telophase

_____ **5.** metaphase

_____ **6.** anaphase

a. Chromosomes move to the center of the cell and line up along the equator.

b. A nuclear envelope forms around the chromatids at each pole.

c. Chromosomes coil up and become visible.

d. The two chromatids move toward opposite poles as the spindle fibers attached to them shorten.

Study the following steps of mitosis. Determine the order in which the steps take place. Write the number of each step in the space provided.

_____ **7.** prophase

_____ **8.** telophase

_____ **9.** metaphase

_____ **10.** anaphase

Complete each statement by underlining the correct term or phrase in the brackets.

11. Cytokinesis begins [before / after] mitosis.

12. During cytokinesis in animal cells, the cell is pinched in half by [the cell wall / a belt of proteins].

Name _____ Class _____ Date _____

Active Reading

Section: Chromosomes
Read the passage below. Then answer the questions that follow.

A gene is a segment of DNA that codes for a protein or RNA molecule. A single molecule of DNA has thousands of genes lined up like the cars of a train. When genes are being used, the strand of DNA is stretched out so that the information it contains can be decoded and used to direct the synthesis of proteins needed by the cell.

As a eukaryotic cell prepares to divide, the DNA and the proteins associated with the DNA coil into a structure called a **chromosome.** Before the DNA coils up, however, the DNA is copied. The two exact copies of DNA that make up each chromosome are called **chromatids.** The two chromatids, which become separated during cell division and are placed into each new cell, ensure that each new cell has the same genetic information as the original cell.

SKILL: READING EFFECTIVELY
Read each question, and write your answer in the space provided.

1. How are genes and DNA related?

_____

_____

2. What occurs to a DNA strand as its genes are being used?

_____

_____

3. How are chromatids and chromosomes related?

_____

An analogy is a comparison. In the space provided, write the letter of the term or phrase that best completes the analogy.

_____ **4.** A train is to cars as a molecule of DNA is to
  a. chromatids.
  b. genes.
  c. proteins.
  d. RNA.

Skills Worksheet

Active Reading

Section: The Cell Cycle

Read the passage below. Then answer the questions that follow.

The **cell cycle** is a repeating sequence of cellular growth and division during the life of an organism. A cell spends 90 percent of its time in the first three phases of the cycle, which are collectively called **interphase.** A cell will enter the last two phases of the cell cycle only if it is about to divide.

The five phases of the cell cycle are as follows:

First growth (G₁)phase: During the G_1 phase, a cell grows rapidly and carries out its routine functions. For most organisms, this phase occupies the major portion of the cell's life.

Synthesis (S) phase: A cell's DNA is copied during this phase. At the end of this phase, each individual chromosome consists of two chromatids attached at the centromere.

Second growth (G₂) phase: In the G_2 phase, preparations are made for the nucleus to divide. Mitochondria and other organelles replicate. Hollow protein fibers called microtubules are assembled. The microtubules are used to move the chromosomes during mitosis.

Mitosis: The process during cell division in which the nucleus of a cell is divided into two nuclei is called **mitosis.** Each nucleus ends up with the same number and kinds of chromosomes.

Cytokinesis: The process during cell division in which the cytoplasm divides is called cytokinesis.

SKILL: READING EFFECTIVELY

Read each question, and write your answer in the space provided.

1. What two key terms are contained in the first paragraph of this passage?

_____

2. Give the meaning of these two terms.

_____

_____

_____

3. A cell viewed under a high-powered microscope appears to be in the fourth phase of the cell cycle. What does this indicate about the cell?

_____

_____

SKILL: SEQUENCING INFORMATION

Match each statement with the phase of the cell cycle it describes. Write the letter of the correct phase in the space provided. Some choices may be used more than once.

_____ **4.** nucleus divides

_____ **5.** makes up a major portion of
       most cells' lives

_____ **6.** cytoplasm divides

_____ **7.** mitochondria replicate

_____ **8.** cell grows rapidly

_____ **9.** two identical nuclei are produced

_____ **10.** DNA is copied

_____ **11.** microtubules are assembled

_____ **12.** forms two chromatids attached at the centromere

_____ **13.** cell carries out its routine functions

_____ **14.** microtubules move chromosomes

a. first growth phase

b. synthesis phase

c. second growth phase

d. mitosis

e. cytokinesis

Read the question, and write your answer in the space provided.

15. How are mitosis and cytokinesis alike? How do they differ?

_____

_____

_____

An analogy is a comparison. In the space provided, write the letter of the term or phrase that best completes the analogy.

_____ **16.** G_2 phase is to mitochondria as S phase is to
    a. chromatids.
    b. centromere.
    c. microtubules.
    d. DNA.

Skills Worksheet

Active Reading

Section: Mitosis and Cytokinesis

Read the passage below. Then answer the questions that follow.

During cytokinesis, the cytoplasm of the cell is divided in half, and the cell membrane grows to enclose each cell, forming two separate cells as a result.

During cytokinesis in animal cells and other cells that lack cell walls, the cell is pinched in half by a belt of protein threads.

Plant cells and other cells that have rigid cell walls have different method of dividing the cytoplasm. In plant cells, vesicles formed by the Golgi apparatus fuse at the midline of the dividing cell and form a cell plate. A cell plate is a membrane-bound cell wall that forms across the middle of the cell. A new cell wall then forms on both sides of the cell plate.

SKILL: READING EFFECTIVELY

In the space provided, match each statement with the stage of cellular division it describes. Write a if the statement describes cytokinesis in animal cells, write p if it describes cytokinesis in plant cells, or write b if it describes cytokinesis in both.

_____ **1.** The Golgi apparatus forms vesicles.

_____ **2.** Two genetically identical cells are formed.

_____ **3.** A belt of protein thread pinches the cell in half.

_____ **4.** A cell plate forms across the cell's middle.

_____ **5.** The cytoplasm of the cell divides in half.

_____ **6.** A cell wall forms on both sides of cell plate.

An analogy is a comparison. In the space provided, write the letter of the term or phrase that best completes the analogy.

_____ **7.** Plant cell is to cell plate as animal cell is to
   a. nucleus.
   b. cytoplasm.
   c. protein threads.
   d. Both (a) and (b)

Name _____ Class _____ Date _____

Vocabulary Review

In the space provided, write the letter of the term or phrase that best completes each statement or best answers each question.

_____ **1.** An organism's reproductive cells, such as sperm or egg cells, are called
    a. genes.
    b. chromosomes.
    c. gametes.
    d. zygotes.

_____ **2.** A form of asexual reproduction in bacteria is
    a. binary fission.
    b. trisomy.
    c. mitosis.
    d. development.

_____ **3.** A segment of DNA that codes for a protein or RNA molecule is a
    a. chromosome.
    b. gene.
    c. chromatid.
    d. centromere.

_____ **4.** At the beginning of cell division, DNA and the proteins associated with the DNA coil into a structure called a(n)
    a. chromatid.
    b. autosome.
    c. centromere.
    d. chromosome.

_____ **5.** The two exact copies of DNA that make up each chromosome are called
    a. homologous chromosomes.
    b. centromeres.
    c. chromatids.
    d. autosomes.

_____ **6.** The two chromatids of a chromosome are attached at a point called the
    a. diploid.
    b. centriole.
    c. spindle.
    d. centromere.

_____ **7.** Chromosomes that are similar in size, shape, and genetic content are called which of the following?
    a. homologous chromosomes
    b. haploid
    c. diploid
    d. karyotypes

_____ **8.** When a cell contains two sets of chromosomes, it is said to be
    a. haploid.
    b. binary.
    c. diploid.
    d. saturated.

_____ **9.** When a cell contains one set of chromosomes, it is said to be
    a. haploid.
    b. separated.
    c. diploid.
    d. homologous.

_____ **10.** The fertilized egg, the first cell of a new individual, is called a(n)
    a. autosome.
    b. zygote.
    c. organism.
    d. chromosome.

| Vocabulary Review *continued*

_____11. A photo of the chromosomes in a dividing cell, arranged by size, is
a(n)
    a. electronic scan.   **c.** X ray.
    b. karyotype.   **d.** anaphase.

_____12. What are chromosomes not directly involved in determining the sex of
an individual?
    a. asexual chromosomes   **c.** autosomes
    b. chromatids   **d.** haploid

_____13. Chromosomes that contain genes that will determine the sex of the
individual are called
    a. X chromosomes.   **c.** Y chromosomes.
    b. sex chromosomes.   **d.** autosomes.

_____14. The repeated sequence of growth and division during the life of a cell
is called the
    a. cell cycle.   **c.** binary fission.
    b. cytokinesis.   **d.** amniocentesis.

_____15. The first three phases of the cell cycle are called
    a. anaphase.   **c.** mitosis.
    b. interphase.   **d.** synthesis phase.

_____16. What is the process during which the nucleus of a cell is divided into
two nuclei?
    a. fertilization   **c.** mitosis
    b. disjunction   **d.** cytokinesis

_____17. The process during cell division in which the cytoplasm divides is
called
    a. cytokinesis.   **c.** interphase.
    b. trisomy.   **d.** mitosis.

_____18. What is the uncontrolled division of cells?
    a. Down syndrome   **c.** cancer
    b. mutation   **d.** trisomy

_____19. Cell structures made of individual microtubule fibers that are involved
in moving chromosomes during cell division are called
    a. chromatids.   **c.** centrioles.
    b. fertilizers.   **d.** spindles.

Skills Worksheet

Science Skills

Sequencing

The figure below illustrates the life cycle of a eukaryotic cell, which is known as the cell cycle. The names of the phases have been omitted from the figure. Use the figure below to complete items 1–8.

In the space provided in the figure below, write the letter of the phase of the cell cycle that matches each phase in the figure.

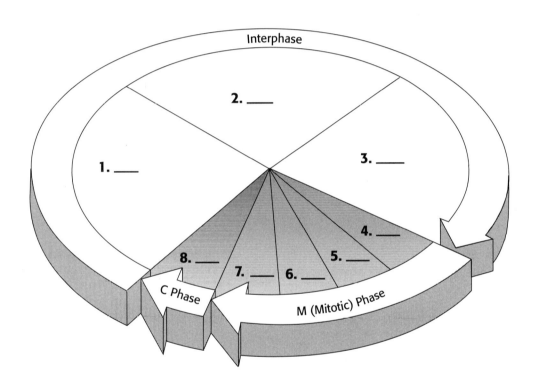

PHASES OF THE CELL CYCLE

a. prophase **e.** S
b. G_1 **f.** cytokinesis
c. telophase **g.** G_2
d. metaphase **h.** anaphase

Science Skills *continued*

In the space provided below each animal cell, write the name of the stage of mitosis that is represented.

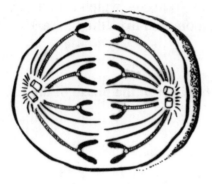

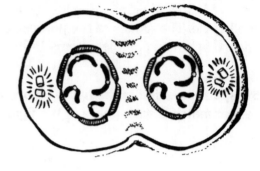

9. _____   10. _____

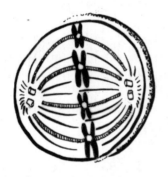

11. _____   12. _____

Determine the order in which the following four stages of mitosis take place. Write the number of each step in the space provided.

_____**13.** anaphase

_____**14.** metaphase

_____**15.** telophase

_____**16.** prophase

Name _____ Class _____ Date _____

Concept Mapping

Using the terms and phrases provided below, complete the concept map showing the principles of cell division.

asexual reproduction   growth   sexual reproduction

cytokinesis   prokaryotes   synthesis phase

eukaryotes   repair

first growth phase   second growth phase

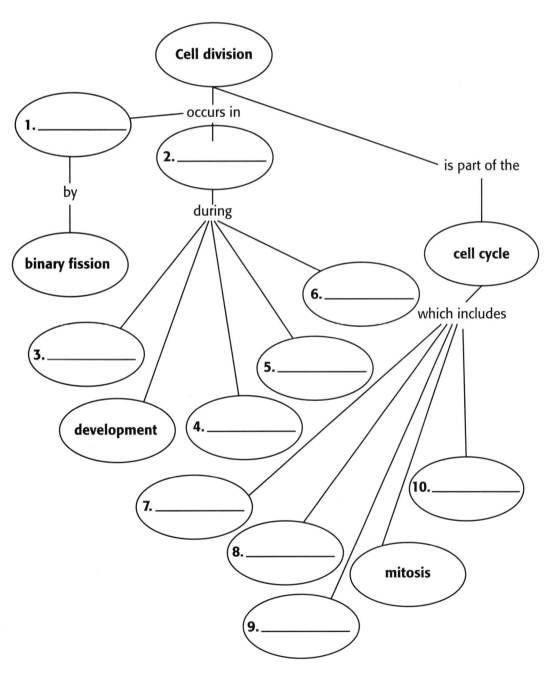

Skills Worksheet

Critical Thinking

Look-Alikes

In the space provided, write the letter of the term or phrase that best describes how each numbered item looks.

_____ **1.** super coil

_____ **2.** chromosome

_____ **3.** metaphase

_____ **4.** genes in a chromosome

_____ **5.** spindle

a. a large, thick spring

b. cars of a train

c. a jump rope with an X on it

d. an X

e. a long line of X's jumping rope

Work-Alikes

In the space provided, write the letter of the term or phrase that best describes how each numbered item functions.

_____ **6.** karyotype

_____ **7.** homologous chromosomes

_____ **8.** stages of mitosis

_____ **9.** haploid cell

_____ **10.** cell cycle

a. twins

b. traffic light

c. scenes of a play

d. chessboard with only the white pieces

e. a family picture

Cause and Effect

In the space provided, write the letter of the term or phrase that best matches each cause or effect given below.

Cause	Effect	
11. _____	start of mitosis	**a.** plant cell separates
12. mitosis	_____	**b.** trisomy 21
13. _____	development of male	**c.** Y chromosome
14. _____	Down syndrome	**d.** nuclear membrane dissolves
15. cell plate forms	_____	**e.** one nucleus becomes two nuclei
16. changes in chromosome structure	_____	**f.** mutations

| Critical Thinking *continued*

Linkages

In the spaces provided, write the letters of the two terms or phrases that are linked together by the term or phrase in the middle. The choices can be placed in any order.

17. _____ mitosis _____

18. _____ synthesis (S) phase _____

19. _____ anaphase _____

20. _____ DNA synthesis (G_2) checkpoint _____

21. _____ protein may not function _____

a. mitosis checkpoint

b. interphase

c. gene mutation

d. second growth (G_2) phase

e. cancer

f. cytokinesis

g. cell growth (G_1) checkpoint

h. telophase

i. first growth (G_1) phase

j. metaphase

Analogies

An analogy is a relationship between two pairs of terms or phrases written as a : b :: c : d. The symbol : is read as "is to," and the symbol :: is read as "as." In the space provided, write the letter of the pair of terms or phrases that best completes the analogy shown.

_____ **22.** chromosome : DNA and protein ::
- **a.** gene : centromere
- **b.** chromatid : gamete
- **c.** spindle fiber : microtubule
- **d.** centromere : chromatid

_____ **23.** homologous pair : chromosomes ::
- **a.** chromatid : chromosomes
- **b.** egg : gametes
- **c.** chromosome: chromatids
- **d.** gamete : zygotes

_____ **24.** prophase : telophase ::
- **a.** prophase : metaphase
- **b.** metaphase : anaphase
- **c.** anaphase : telophase
- **d.** telophase : metaphase

_____ **25.** metaphase : meet ::
- **a.** anaphase : separate
- **b.** prophase : double
- **c.** telophase : double
- **d.** telophase : lineup

Name _____ Class _____ Date _____

Test Prep Pretest

In the space provided, write the letter of the term or phrase that best completes each statement or best answers each question.

_____ **1.** As a cell prepares to divide, a DNA molecule and its associated proteins coil to form a
   a. chromatid.
   b. gene.
   c. chromosome.
   d. centromere.

_____ **2.** What is the number of chromosomes found in a human body cell?
   a. 23
   b. 46
   c. 48
   d. 64

_____ **3.** The sex of a human offspring is determined by
   a. the female.
   b. the male.
   c. both the female and the male.
   d. neither the female nor the male.

_____ **4.** Bacteria reproduce through an asexual process called
   a. meiosis.
   b. cytokinesis.
   c. interphase.
   d. binary fission.

_____ **5.** In plant cells, cytokinesis requires the formation of a new
   a. Golgi apparatus.
   b. cell wall.
   c. centromere.
   d. series of protein threads.

_____ **6.** Gene mutations that result in cancer often cause the
   a. overproduction of growth-promoting proteins.
   b. underproduction of growth-promoting proteins.
   c. activation of control proteins that slow or stop the cell cycle.
   d. Both (a) and (c)

_____ **7.** Which of the following is NOT part of the spindle apparatus in animal cells?
   a. microtubules
   b. belt of protein threads
   c. spindle fibers
   d. centrioles

Complete each statement by writing the correct term or phrase in the space provided.

8. A(n) _____ is a segment of DNA that transmits information from parent to offspring.

9. An individual with an extra copy of chromosome 21 demonstrates traits collectively known as _____ _____ .

10. The 22 pairs of chromosomes in human somatic cells that are the same in males and females are called _____ .

11. The human chromosomes that determine an individual's sex are called the

_____ _____ .

Questions 12–17 refer to the sequence below.

$$G_1 \longrightarrow S \longrightarrow G_2 \longrightarrow M \longrightarrow C$$

12. The sequence above represents the _____

_____ .

13. The S in the sequence represents the phase in which

_____ _____ occurs.

14. Phases G_1, S, and G_2 in the sequence above are collectively called

_____.

15. Each individual protein structure that helps to move the chromosomes apart during mitosis is called a(n) _____ .

16. A disease caused by uncontrolled cell division is _____ .

17. In the first stage of binary fission, the DNA is _____ .

Read each question, and write your answer in the space provided.

18. What happens to the structure of DNA in your cells prior to cell division?

_____

_____

_____

19. Explain the difference in the number of chromosomes between a frog somatic cell and a frog egg cell.

_____

_____

_____

20. What happens when nondisjunction takes place during cell division?

_____

_____

_____

_____

_____

21. Describe what happens at each checkpoint during the cell cycle.

_____

_____

_____

_____

_____

_____

22. What are the four stages of mitosis in the correct order?

_____

_____

23. Explain the events that take place during each stage of mitosis.

_____

_____

_____

_____

_____

_____

_____

Name _____ Class _____ Date _____

| Test Prep Pretest *continued*

24. Describe the events that take place during each phase of interphase.

_____

_____

_____

_____

_____

_____

_____

25. List four types of events that take place in a eukaryotic organism, such as a deer, that require cell division.

_____

_____

_____

_____

Name _____ Class _____ Date _____

Quiz

Section: Chromosomes

In the space provided, write the letter of the description that best matches the term or phrase.

_____ **1.** zygote

_____ **2.** DNA

_____ **3.** karyotype

_____ **4.** mutations

_____ **5.** gametes

a. egg cells and sperm cells

b. supplies information that directs a cell's activities and determines its characteristics

c. changes in an organism's genetic material

d. a diploid cell that results from the fusion of two haploid gametes

e. a picture of the chromosomes found in an individual's cells

In the space provided, write the letter of the term or phrase that best completes each statement or best answers each question.

_____ **6.** In human sexual reproduction, a male haploid gamete and a female haploid gamete unite to form which of the following?
a. an egg cell with 46 chromosomes
b. a zygote with 23 chromosomes
c. a zygote with 46 chromosomes
d. a sperm cell with 23 chromosomes

_____ **7.** Chromosomes that determine the sex of an individual are called
a. autosomes.
b. sex chromosomes.
c. homologous chromosomes.
d. chromatids.

_____ **8.** Chromosomes that are similar in size, shape, and genetic content are called
a. autosomes.
b. sex chromosomes.
c. homologous chromosomes.
d. chromatids.

_____ **9.** If nondisjunction occurs, one of the resulting cells will receive
a. two homologues of a chromosome.
b. no homologues of a chromosome.
c. three homologues of a chromosome.
d. one homologue of a chromosome.

_____ **10.** A mutation in which the chromosome piece reattaches to the original chromosome but in the reverse orientation, is known as a(n)
a. duplication.
b. inversion.
c. deletion.
d. translocation.

Assessment

Quiz

Section: The Cell Cycle

In the space provided, write the letter of the description that best matches the term or phrase.

_____ 1. cytokinesis

_____ 2. cell cycle

_____ 3. mitosis

_____ 4. synthesis (S) phase

_____ 5. first growth (G₁) phase

a. chromatids are formed

b. nucleus is divided into two nuclei

c. major portion of the cell's life

d. cytoplasm divides

e. a repeating sequence of cellular growth and division

In the space provided, write the letter of the term or phrase that best completes each statement or best answers each question.

_____ 6. Cells that might never divide include which of the following?
    a. skin cells
    b. nerve cells
    c. sex cells
    d. All of the above

_____ 7. The cell cycle controls cell division
    a. only in eukaryotes.
    b. only in prokaryotes.
    c. in both eukaryotes and prokaryotes.
    d. only in multicellular organisms.

_____ 8. Mitosis occurs
    a. immediately after the synthesis phase.
    b. before the second growth phase.
    c. after the second growth phase.
    d. after cytokinesis.

_____ 9. Which of the following is the LEAST likely cause of cancer?
    a. overproduction of growth-promoting proteins
    b. overproduction of control proteins that slow the cell cycle
    c. inactivation of control proteins that slow the cell cycle
    d. inactivation of control proteins that stop the cell cycle

_____ 10. DNA replication is checked during the
    a. cell growth (G₁) checkpoint.
    b. DNA synthesis (G₂) checkpoint.
    c. mitosis checkpoint.
    d. cytokinesis checkpoint.

Assessment

Quiz

Section: Mitosis and Cytokinesis

In the space provided, write the letter of the description that best matches the term or phrase.

_____ 1. centriole

_____ 2. spindle fiber

_____ 3. anaphase

_____ 4. cell plate

_____ 5. prophase

a. made of individual microtubules

b. nuclear envelope dissolves

c. membrane-bound cell wall

d. centromeres divide

e. made of nine triplets of microtubules

In the space provided, write the letter of the term or phrase that best completes each statement or best answers each question.

_____ 6. Which of the following is NOT a stage of mitosis?
  a. prophase
  b. metaphase
  c. cytokinesis
  d. telophase

_____ 7. Which of the following does NOT play a role in cytokinesis in plant cells?
  a. Golgi apparatus
  b. belt of protein threads
  c. cell membrane
  d. cell plate

_____ 8. During telophase, the
  a. cytoplasm divides.
  b. nuclear membrane dissolves.
  c. chromosomes line up in the center of the cell.
  d. None of the above

_____ 9. Plant cells
  a. lack centrioles.
  b. lack a cell membrane.
  c. divide to produce cells of unequal size.
  d. All of the above

_____ 10. Cell division in eukaryotes differs from cell division in prokaryotes because
  a. in eukaryotes, the organelles must also be divided between cells.
  b. prokaryotes lack a nucleus.
  c. prokaryotes have a circular chromosome.
  d. All of the above

Name _____ Class _____ Date _____

Chapter Test

Chromosomes and Cell Reproduction

In the space provided, write the letter of the description that best matches the term or phrase.

_____ 1. chromatids

_____ 2. centromere

_____ 3. homologous chromosomes

_____ 4. diploid

_____ 5. haploid

_____ 6. deletion

_____ 7. cancer

_____ 8. first growth (G_1) phase

_____ 9. mitosis checkpoint

_____ 10. second growth (G_2) phase

a. a cell that contains one set of chromosomes

b. preparations are made for the nucleus to divide

c. a mutation that occurs when a chromosome fragment breaks off and is lost

d. misregulation of the proteins that control cell growth and division

e. the point at which two chromatids are attached

f. the two copies of DNA on each chromosome that form just before cell division

g. the cell grows and carries out routine functions

h. chromosomes that are similar in shape and size and have similar genetic information

i. a cell that contains two sets of chromosomes

j. triggers the exit from mitosis

In the space provided, write the letter of the term or phrase that best completes each statement or best answers each question.

_____ 11. A student can study a karyotype to learn about the
  a. process of binary fission.
  b. genes that are present in a particular strand of DNA.
  c. medical history of an individual.
  d. chromosomes present in the somatic cell.

_____ 12. Spindles are composed of which of the following?
  a. nine triplets of microtubules
  b. individual microtubule fibers and centrioles
  c. chromatids and centromeres
  d. microtubules arranged in a circle around the centriole

_____ 13. The condition in which a diploid cell has an extra chromosome is called
  a. monosomy.
  b. disjunction.
  c. trisomy.
  d. karyotype.

Name _____ Class _____ Date _____

Questions 14 and 15 refer to the figure below, which shows the stages of mitosis.

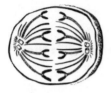   

  A   B   C   D

_____**14.** Which of the following correctly indicates the order in which mitosis occurs?
 a. A, B, C, D
 b. B, A, C, D
 c. C, B, A, D
 d. A, C, B, D

_____**15.** Which stage shows metaphase?
 a. A
 b. B
 c. C
 d. D

Questions 16 and 17 refer to the figures below.

Animal Cell   **Plant Cell**

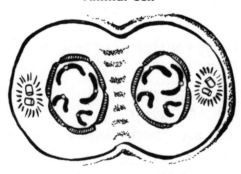 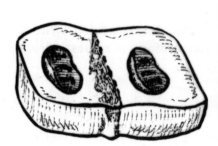

_____**16.** The stage of the cell cycle that these cells are in is
 a. first growth (G$_1$) phase.   **c.** mitosis.
 b. synthesis (S) phase.   **d.** cytokinesis.

_____**17.** The structure in the center of the animal cell that pinches the cell in half is called the
 a. belt of protein threads.
 b. Golgi apparatus.
 c. centromere.
 d. spindle apparatus.

Chapter Test *continued*

_____**18.** The structure in the center of the plant cell could NOT be formed without which of the following?
　　a. cell plate
　　b. Golgi apparatus
　　c. cell membrane
　　d. All of the above

_____**19.** The repeating sequence of growth and division through which many eukaryotic cells pass is called
　　a. the cell cycle.
　　b. binary fission.
　　c. cytokinesis.
　　d. meiosis.

_____**20.** Eukaryotes use cell division to
　　a. grow.
　　b. develop.
　　c. reproduce.
　　d. All of the above

Assessment

Chapter Test

Chromosomes and Cell Reproduction

In the space provided, write the letter of the description that best matches the term or phrase.

_____ **1.** cell growth (G_1) checkpoint

_____ **2.** anaphase

_____ **3.** DNA synthesis (G_2) checkpoint

_____ **4.** prophase

_____ **5.** metaphase

a. chromatids move to opposite poles of the cell

b. chromosomes line up at the equator of the cell

c. determines whether a cell will divide

d. the nuclear membrane dissolves and chromosomes become visible

e. DNA repair enzymes check DNA replication

In the space provided, write the letter of the term or phrase that best completes each statement or best answers each question.

_____ **6.** Eukaryotes and prokaryotes both use cell division to
  a. grow.
  b. repair damage.
  c. reproduce.
  d. All of the above

_____ **7.** Chromatids are
  a. made of microtubules.
  b. bacterial chromosomes.
  c. strands of duplicate genetic material.
  d. supercoils of protein.

_____ **8.** If conditions are favorable for cell division during the G_1 phase,
  a. the microtubules attach to the centromeres.
  b. the nucleus begins to divide.
  c. the centriole pair is replicated.
  d. proteins stimulate the cell to copy its DNA.

_____ **9.** If nondisjunction occurs,
  a. too many gametes will be produced.
  b. no gametes will be produced.
  c. a gamete will receive too many or too few homologues of a chromosome.
  d. a mutation occurs.

_____ **10.** Which stage of the cell cycle occupies most of the cell's life?
  a. G_1
  b. M
  c. G_2
  d. S

Complete each statement by writing the correct term or phrase in the space provided.

11. A fertilized egg cell is called a(n) _____ .

12. The process by which a bacterial cell splits asexually into two identical organ-

isms is called _____ _____ .

13. Collectively, the time spent in G_1, S, and G_2 is called

_____ .

14. A zygote is formed by the union of two haploid _____
from the opposite sexes.

15. During cell division, the movement of chromosomes is aided by a structure

called the _____ .

16. Cancer can occur when _____ mutate and the

_____ that regulate cell growth and division do not

function correctly.

17. Chromosomes that are similar in shape, size, and genetic content are called

_____ _____ .

18. A DNA molecule contains thousands of _____ , and the

DNA and its associated proteins form _____ .

19. If a piece of a chromosome breaks off, a(n) _____

_____ occurs.

20. The information a cell needs to direct its activities and to determine its char-

acteristics is contained in molecules of _____ .

Read each question, and write your answer in the space provided.

21. Describe the role that the spindle apparatus plays in mitosis.

_____

_____

_____

22. Describe how a karyotype can be used to diagnose Down syndrome.

_____

_____

_____

_____

23. Describe the major events that occur during each of the five stages of the cell cycle.

_____

_____

_____

_____

_____

_____

_____

24. Describe the difference between cytokinesis in animal cells and in plant cells.

_____

_____

_____

_____

_____

25. Explain what might happen if the proteins that normally act to slow or stop the cell cycle are inactivated.

_____

_____

_____

_____

Name _____ Class _____ Date _____

Modeling Chromosomal Mutations

You can use paper and a pencil to model the ways in which chromosome structure can change.

MATERIALS

- 14 note-card pieces
- pencils
- tape

Procedure

1. Write the numbers 1–8 on note-card pieces (one number per piece). Tape the pieces together in numerical order to model a chromosome with eight genes.

2. Use the "chromosome" you made to model the four alterations in chromosome structure illustrated below. For example, remove the number 3 and reconnect the remaining chromosome pieces to represent a deletion.

3. Reconstruct the original chromosome before modeling a duplication, an inversion, and a translocation. Use the extra note-card pieces to make the additional numbers you need.

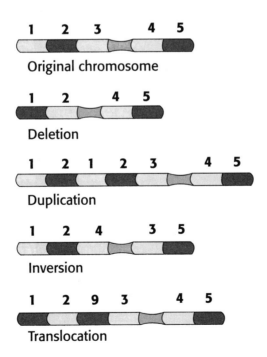

1 2 3 4 5
Original chromosome

1 2 4 5
Deletion

1 2 1 2 3 4 5
Duplication

1 2 4 3 5
Inversion

1 2 9 3 4 5
Translocation

Modeling Chromosomal Mutations *continued*

Analysis

Describe how a cell might be affected by each mutation if the cell were to receive a chromosome with that mutation.

_____

_____

_____

_____

_____

_____

_____

Name _____ Class _____ Date _____

DATASHEET FOR IN-TEXT LAB

Calculating the Number of Cells Resulting from Mitosis

Background

Scientists investigating cancer might need to know the number of cells produced in a certain amount of time. In the human body the rate of mitosis is about 25 million (2.5×10^7) cells produced every second! You can calculate the number of cells produced by mitosis in a given amount of time.

1. **Calculate the number of cells produced by mitosis in the given time.** For example, to find the number of cells produced in 3 minutes, determine how many seconds there are in 3 minutes (since the rate is given in seconds).

$$\frac{60 \text{ seconds} \times 3 \text{ minutes}}{1 \text{ minute}} = 180 \text{ seconds}$$

2. **Multiply the rate of mitosis by the time (in seconds) asked for in the problem (180 seconds).**

$$\frac{2.5 \times 10^7 \text{ cells} \times 180 \text{ seconds}}{\text{second}} = 4.5 \times 10^9 \text{ cells} (4,500,000,000 \text{ cells})$$

Analysis

1. **Calculate** the number of cells that would be produced in 1 hour.

_____

2. **Calculate** the number of cells that would be produced in 1 day.

_____

3. **Critical Thinking**
   Predicting Patterns Identify factors that might increase or decrease the rate of mitosis.

_____

_____

_____

Quick Lab

Observing Mitosis and Cytokinesis

You can identify the stages of mitosis and the process of cytokinesis by observing slides of tissues undergoing mitosis using a compound microscope.

MATERIALS

- compound microscope
- prepared slide of mitosis
- paper
- pencil

Procedure

1. View a prepared slide of cells undergoing mitosis under low power of a compound microscope.

2. Move the slide until you find a section where different stages of mitosis are visible.

3. Switch to high power. Use photos or diagrams from your textbook to help you locate and identify cells in interphase and in each stage of mitosis.

4. On a separate piece of paper, sketch an example of each stage. Label each sketch with the following terms where appropriate: *chromosomes, cell membrane, cytoplasm, nucleus, spindle,* and *cell wall.*

5. Switch to low power, and estimate how many cells are clearly in interphase and how many cells are in one of the stages of mitosis.

Analysis

1. **Describe** the activity of chromosomes in each stage of mitosis.

_____

_____

_____

_____

_____

2. **Compare** the number of cells in interphase with the number of cells in one of the stages of mitosis.

_____

_____

_____

Observing Mitosis and Cytokinesis *continued*

3. Critical Thinking
Inferring Relationships What does your answer to item 2 indicate about the relative length of interphase?

_____

_____

Exploration Lab)   **DATASHEET FOR IN-TEXT LAB**

Modeling Mitosis

SKILLS

• Modeling

• Using scientific methods

OBJECTIVES

• **Describe** the events that occur in each stage of mitosis.

• **Relate** mitosis to genetic continuity.

MATERIALS

• pipe cleaners of at least two different colors

• yarn

• wooden beads

• white labels

• scissors

Before You Begin

The cell cycle includes all of the phases in the life of a cell. The **cell cycle** is a repeating sequence of cellular growth and division during the life of an organism. Mitosis is one of the phases in the cell cycle. **Mitosis** is the process by which the material in a cell's nucleus is divided during cell reproduction. In this lab, you will build a model that will help you understand the events of mitosis. You can also use the model to demonstrate the effects of **nondisjunction** and **mutations.**

1. Write a definition for each boldface term in the paragraph above and for the following terms: chromatid, centromere, spindle fiber, cytokinesis. Use a separate sheet of paper.

2. Where in the human body do cells undergo mitosis?

   _____

3. How does a cell prepare to divide during interphase of the cell cycle?

   _____

   _____

4. Based on the objectives for this lab, write a question you would like to explore about mitosis.

   _____

   _____

Modeling Mitosis *continued*

Procedure

PART A: DESIGN A MODEL

1. Work with the members of your lab group to design a model of a cell that uses the materials listed for this lab. Be sure your model cell has at least two pairs of chromosomes and is about to undergo mitosis.

You Choose

As you design your model, decide the following:

  a. what question you will explore

  b. how to construct a cell membrane

  c. how to show that your cell is diploid

  d. how to show the locations of at least two genes on each chromosome

  e. how to show that chromosomes are duplicated before mitosis begins

---

2. Write out the plan for building your model. Have your teacher approve the plan before you begin building the model.

_____

_____

_____

_____

3. Build the cell model your group designed. **CAUTION: Sharp or pointed objects can cause injury. Handle scissors carefully. Promptly notify your teacher of any injuries.** Use your model to demonstrate the phases of mitosis. Draw and label each phase you model.

4. Use your model to explore one of the questions written for step 4 of **Before You Begin.** Describe the steps you took to explore the question.

_____

_____

_____

_____

_____

| Modeling Mitosis *continued*

PART B: TEST HYPOTHESES

Answer each of the following questions by writing a hypothesis. Use your model to test each hypothesis, and describe your results.

5. Cytokinesis follows mitosis. How will the size of each new cell that is formed following cytokinesis compare with that of the original cell?

_____

_____

6. Sometimes two chromatids fail to separate during mitosis. How might this failure affect the chromosome number of the two new cells?

_____

_____

_____

7. A mutation is a permanent change in a gene or chromosome. What effect might a mutation in a parent cell have on future generations of cells that result from the parent cell?

_____

_____

PART C: CLEANUP AND DISPOSAL

8. Dispose of paper and yarn scraps in the designated waste container.

9. Clean up your work area and all lab equipment. Return lab equipment to its proper place. Wash your hands thoroughly before you leave the lab and after you finish all work.

Analyze and Conclude

1. Analyzing Results How do the nuclei you made by modeling mitosis compare with the nucleus of the model cell you started with? Explain your result.

_____

_____

_____

_____

| Modeling Mitosis *continued*

   2. Evaluating Methods How could you modify your model to better illustrate the process of mitosis?

_____

_____

_____

_____

   3. Recognizing Patterns How does the genetic makeup of the cells that result from mitosis compare with the genetic makeup of the original cell?

_____

_____

_____

_____

   4. Inferring Conclusions How is mitosis important?

_____

_____

_____

   5. Further Inquiry Write a new question about mitosis or the cell cycle that could be explored with your model.

_____

_____

Name _____ Class _____ Date _____

Mitosis

All cells undergo a process of growth and division called the *cell cycle*. The cell cycle consists of three major stages: *interphase*, *mitosis*, and *cytokinesis*. During interphase, the cell grows and the cell's DNA replicates. The next phase is mitosis, during which the replicated genetic material separates into two identical nuclei. Mitosis is divided into four stages: *prophase*, *metaphase*, *anaphase*, and *telophase*. Cytokinesis, the last stage of cell division, is the division of the cell cytoplasm between the two new, genetically identical daughter cells.

In this lab, you will study and compare mitosis in animal cells and plant cells. You will also identify cells in the different stages of mitosis.

OBJECTIVES

Observe and **identify** cells undergoing the process of mitosis.

Compare the stages of mitosis in plant cells with these stages in animal cells.

MATERIALS

- compound light microscope
- prepared slides (longitudinal sections of onion or garlic root tips)
- prepared slides (whitefish blastula or other animal tissue)

Procedure

PART 1: MITOSIS IN PLANT CELLS

1. Observe a prepared slide of a longitudinal section of onion or garlic root tips. Use low power to locate the region of actively dividing cells near the end of the root. Using high power, examine individual cells in the region. Find a cell from each stage of mitosis, using the descriptions below. In the space next to each description, draw a cell in that stage as it appears on your slide. Label the visible part of the cell.

Interphase

This is the phase of normal cell activity. During interphase, individual chromosomes cannot be distinguished. Instead they appear as a dark mass of material called *chromatin*. The DNA of each chromosome replicates at the end of this stage. Note the nucleus with one or more dark-stained nucleoli filled with chromatin.

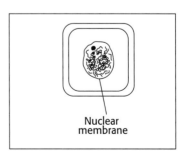

Nuclear membrane

Prophase

The chromatin appears as a mass of thick threads. These threads are the replicated chromosomes, which have coiled up and shortened. Each chromosome consists of a pair of *chromatids*, which are duplicates of the original chromosome. The chromatids are held together by a centromere. In late prophase, the chromosomes are distinctly visible as pairs of chromatids in the central region of the cell.

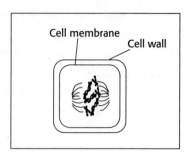

Metaphase

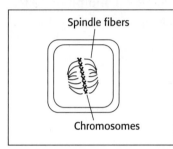

The chromosomes line up across the equator of the cell. A mass of fibers called a *spindle* has formed between the poles of the cell and the mass of chromosomes. A spindle fiber from each pole attaches to each pair of chromatids.

Anaphase

The centromere of each chromatid pair divides. The chromatids move along the spindle fibers toward the poles of the cell. Each chromatid in the pair of chromatids moves toward opposite poles of the cell.

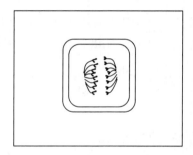

Telophase

The chromatids (now called chromosomes) have formed clumps at each pole. A new nuclear membrane forms around the chromosomes, which uncoil and return to the chromatin network seen in interphase. In plants, the new cell walls grow to form the two new, identical daughter cells.

2. Describe the shape of the cells and the color of the chromosomes.

_____

_____

Mitosis *continued*

PART 2: MITOSIS IN ANIMAL CELLS

3. Observe the prepared slide of the whitefish blastula or other animal cells. Use low power to locate the cells, then examine them under high power. Find a cell from each stage, using the descriptions below. In the space near each description, draw a cell in that stage as it appears on your slide. Label the visible parts of the cell.

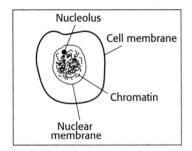

Interphase

A distinct nucleus and a nucleolus are visible. The genetic material appears as chromatin.

Prophase

In early prophase, astral rays have formed around the centrioles, and the spindle has formed between them. The paired chromatids are becoming visible, and the nuclear membrane has disappeared. In late prophase, the chromosomes are short and thick, and are distinct in the central region of the cell.

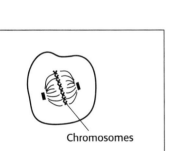

Metaphase

The chromosomes line up at the center of the cell, along the equator.

Anaphase

The chromatids separate at their centromere and are pulled to opposite poles along the spindle fibers.

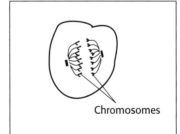

Telophase

The chromosomes appear in clusters at the poles. The parent cell begins to elongate, and the nuclear membranes reform around the chromosome clusters. The spindle and chromosomes become less distinct. The cytoplasm pinches in until the two daughter cells separate during late telophase.

4. Describe the shape of the chromosomes as they are pulled to the poles in anaphase.

_____

5. Clean up your work area and wash your hands before leaving the lab.

Analysis

1. Identifying Relationships How does mitosis differ in plant cells and animal cells?

_____

_____

_____

_____

_____

2. Analyzing Data Which phase of mitosis shows the greatest difference between animal cells and plant cells? Explain your choice.

_____

_____

_____

Conclusions

1. Drawing Conclusions What role do you think mitosis plays in living things? Justify your answer.

_____

_____

Extensions

1. Building Models Mitosis in onion root tip cells takes about 80 minutes. If you view a slide of a root tip and count the number of cells in each stage of mitosis, you can then calculate the amount of time each stage takes. This is because the percentage of the cells in a particular stage of mitosis is equal to the percentage of 80 minutes that the stage takes. Using this information, devise a method for calculating the amount of time each stage of mitosis takes.

2. Research and Communications Research the length of time mitosis takes in plant cells and in animal cells.

Skills Practice Lab   **CONSUMER**
Preparing a Root Tip Squash

To observe mitosis in stem and root meristems, biologists prepare a special slide called a *squash*. This preparation is just what it sounds like. Actively dividing cells from a root or stem meristem are removed and treated with hydrochloric acid to fix the cells or to stop them from dividing. The cells are then stained, made into a wet mount, and squashed. Squashing spreads the cells into a single layer.

In this lab, you will study mitotic activity in the roots of onion plants to determine the effects of fertilizer on actively dividing root cells. You also will determine the percentage of cells that are undergoing mitosis. To observe mitosis in the onion root cells, you will make a squash of onion root tips, which contain actively dividing cells.

OBJECTIVES

Prepare and stain slides of onion root tips.

Observe cells in the process of mitosis.

Form and **test** a hypothesis about the effects of fertilizer on mitosis.

Determine the effect of fertilizer on the percentage of cells in mitosis.

MATERIALS

- aceto-orcein stain in dropping bottle (30 mL)
- coverslips (4)
- compound light microscope
- distilled water
- eyedropper
- forceps
- gloves
- 1 M HCl in dropping bottle (6 mL)
- lab apron
- microscope slides (4)
- onion root tips, specially grown
- paper (2 sheets)
- paper towels
- petri dish
- safety goggles
- wooden macerating stick

Procedure

1. Write a hypothesis that clearly states how you think the concentration of fertilizer affects mitosis in onion roots.

_____

_____

PREPARING THE SQUASHES

2. Put on safety goggles, gloves, and a lab apron.

3. Use forceps to carefully remove three onion root tips from the vial labeled "Y—100%," and place them in a petri dish.

4. Use an eyedropper to flood the root tips with 1 M HCl. Allow the root tips to stand in the HCl for 10 minutes. **CAUTION: HCl is an irritant and a poison. Avoid skin and eye contact. If you get HCl on your skin or clothing, wash it off at the sink while notifying your teacher. If you get HCl in your eyes, promptly flush it out at the eyewash station while notifying your teacher. Notify your teacher in the event of an acid spill.**

5. Use the eyedropper to remove the HCl from the petri dish, and dispose of it as instructed by your teacher. Be careful to not remove the root tips with the eyedropper. Refill the petri dish with distilled water.

6. Place a microscope slide on a paper towel. Add three drops of aceto-orcein stain to the center of the slide. **CAUTION: Aceto-orcein stain is corrosive and poisonous. Avoid eye and skin contact. In case of contact, notify your teacher immediately. Avoid inhaling the vapors. Aceto-orcein stain will also stain your skin and clothing. Promptly wash off spills to minimize staining.**

7. Use forceps to transfer a prepared root tip from the petri dish to the drop of stain on the microscope slide.

8. Pulverize the tissue by gently but firmly tapping the root tip with the end of a wooden macerating stick. *Note: Move the stick in a straight up-and-down motion.*

9. Allow the root tip to stain for 10–15 minutes. *Note: Do not let the stain dry. Add more stain if necessary.*

10. Place the slide on a smooth, flat surface. Add a coverslip to the slide to make a wet mount. Place the wet mount between two pieces of paper towel.

11. Use the eraser end of a pencil to press down on the coverslip. Apply only enough pressure to squash the root tip into a single cell layer. Be very careful not to move the coverslip while you are pressing down with the pencil. Do not press too hard because you might break the glass slide or tear apart the cells.

12. Repeat steps 3 through 11 for the C—Control root tips. Place each squash on a piece of paper labeled "C—Control."

| **Preparing a Root Tip Squash** *continued*

OBSERVING THE SQUASHES

13. View one slide at a time with a compound light microscope under both low and high power. *Note: Remember that your mount is fairly thick, so be careful not to change to the high-power objective too quickly. Doing so could shatter the coverslip and destroy your preparation. You will need to focus carefully with the fine-adjustment knob to see the structures under study.* View all three slides you have made. Select the slide that shows the most cells undergoing mitosis. Use this slide to complete steps 14 and 15.
  • What is the shape of the onion root tip cells?

  _____

  • What color did the aceto-orcein make the chromosomes?

  _____

14. Observe the slide under high power. Without moving the slide, estimate the number of cells in the viewing area. To estimate, mentally divide the viewing area into three viewing sections. Then count the cells in one section and record the number for each viewing section in **Table 1.** Total the three numbers and record in the table.

15. In each viewing section, count the number of cells in prophase, metaphase, anaphase, and telophase. Record the numbers in **Table 1.**

TABLE 1 STAGES OF MITOSIS

	Viewing section							
	1		**2**		**3**			
	Y	**C**	**Y**	**C**	**Y**	**C**	**Total Y**	**Total C**
Total number of cells								
							Total number of cells per stage	
Prophase								
Metaphase								
Anaphase								
Telophase								

| Preparing a Root Tip Squash *continued*

• What are the most common stages of mitosis you observed?

_____

_____

16. For the next observation, move the slide to a new viewing section next to the one you just looked at. Repeat steps 14 and 15 in this new section. Repeat steps 14 and 15 for the third viewing section.

17. Repeat steps 14–16 for the C-Control slide you have prepared.

18. Dispose of your materials according to your teacher's instructions. Clean up your work area, and wash your hands before leaving the lab.

Analysis

1. Explaining Events Why do specimens have to be thin to be viewed through the microscope?

_____

_____

2. Describing Events How many cells were in mitosis in each of the slides?

_____

_____

3. Analyzing Data What percentage of cells on your slide were in the process of mitosis? Use the following formula to calculate your answer. Show your calculations in the space provided.

$$\% \text{ cells in mitosis} = \frac{\text{total number of cells in all phases of mitosis}}{\text{total number of cells}} \times 100$$

4. Analyzing Data What percentage of cells on each slide were in each phase of mitosis? Use the following formula to calculate your answer.

$$\% \text{ cells in phase} = \frac{\text{total number of cells in phase}}{\text{total number of cells}} \times 100$$

5. Explaining Events Why did you make two slides in this lab?

_____

_____

_____

Conclusions

1. Evaluating Methods Why do you squash and spread out the root tip?

_____

_____

2. Interpreting Information How do you explain the low percentage of cells undergoing mitosis in each slide?

_____

_____

3. Drawing Conclusions Which slide showed the greatest number of cells in mitosis?

_____

_____

4. Evaluating Methods Why is a root tip a good choice for studying the effects of fertilizer on mitosis?

_____

_____

_____

_____

5. Defending Conclusions What can you conclude about the effect of fertilizer on the growth of an onion plant?

_____

_____

6. Making Predictions Consider the statement "If some is good, more is better." Predict a possible result if you were to double the concentration of fertilizer solution.

_____

_____

_____

7. Evaluating Results Was your hypothesis in step 1 supported? Explain.

_____

_____

_____

Extensions

1. Designing Experiments Design an experiment to test the effect of different formulas of fertilizer on the percentage of cells undergoing mitosis in an onion root tip. For example, identical concentrations of fertilizer could be tested by using different formulas of fertilizer such as 15-30-15, 10-16-16, 30-10-10, or nitrogen only. Be sure to control all variables other than the fertilizer.

2. Research and Communications Botanists are biologists who study plants. They study plant structure, function, growth, and reproduction. Because there are so many things to know about so many different kinds of plants, many botanists specialize in studying just one function of just one kind of plant. Find out about the training and skills required to become a botanist.

Name _____ Class _____ Date _____

Quick Lab

Modeling Chromosomal Mutations

You can use paper and a pencil to model the ways in which chromosome structure can change.

MATERIALS

- 14 note-card pieces
- pencils
- tape

Procedure

1. Write the numbers 1–8 on note-card pieces (one number per piece). Tape the pieces together in numerical order to model a chromosome with eight genes.

2. Use the "chromosome" you made to model the four alterations in chromosome structure illustrated below. For example, remove the number 3 and reconnect the remaining chromosome pieces to represent a deletion.

3. Reconstruct the original chromosome before modeling a duplication, an inversion, and a translocation. Use the extra note-card pieces to make the additional numbers you need.

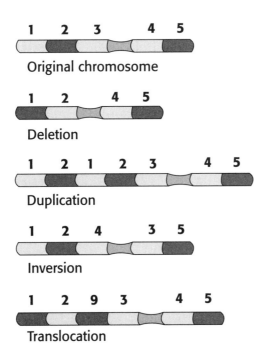

1   2   3   4   5
Original chromosome

1   2   4   5
Deletion

1   2   1   2   3   4   5
Duplication

1   2   4   3   5
Inversion

1   2   9   3   4   5
Translocation

Name _____ Class _____ Date _____

Modeling Chromosomal Mutations *continued*

Analysis

Describe how a cell might be affected by each mutation if the cell were to receive a chromosome with that mutation.

Answers will vary based on the type of mutation: deletion: cell would be

missing a gene, which could prove fatal; duplication: cell would have an

extra gene, which could prove fatal or result in malfunctioning of the cell;

inversion: cell may not be able to use the gene because it is located in a

different area on the chromosome, which could prove fatal; translocation:

cell may not be able to use the gene because it is located on a different

chromosome, which could prove fatal.

Name _____ Class _____ Date _____

Math Lab   **DATASHEET FOR IN-TEXT LAB**

Calculating the Number of Cells Resulting from Mitosis

Background

Scientists investigating cancer might need to know the number of cells produced in a certain amount of time. In the human body the rate of mitosis is about 25 million (2.5×10^7) cells produced every second! You can calculate the number of cells produced by mitosis in a given amount of time.

1. **Calculate the number of cells produced by mitosis in the given time.** For

$$\frac{60 \text{ seconds} \times 3 \text{ minutes}}{1 \text{ minute}} = 180 \text{ seconds}$$

example, to find the number of cells produced in 3 minutes, determine how many seconds there are in 3 minutes (since the rate is given in seconds).

$$\frac{2.5 \times 10^7 \text{ cells} \times 180 \text{ seconds}}{\text{second}} = 4.5 \times 10^9 \text{ cells } (4,500,000,000 \text{ cells})$$

2. **Multiply the rate of mitosis by the time (in seconds) asked for in the problem (180 seconds).**

Analysis

1. **Calculate** the number of cells that would be produced in 1 hour.

   9.0×1010 cells $= 90,000,000,000$ cells

2. **Calculate** the number of cells that would be produced in 1 day.

   2.16×1012 cells

3. **Critical Thinking**
   Predicting Patterns Identify factors that might increase or decrease the rate of mitosis.

   Factors that might increase or decrease the rate of mitosis include mutated

   genes, diet, and exposure to ultraviolet light and tobacco products.

Name _____ Class _____ Date _____

Quick Lab   **DATASHEET FOR IN-TEXT LAB**

Observing Mitosis and Cytokinesis

You can identify the stages of mitosis and the process of cytokinesis by observing slides of tissues undergoing mitosis using a compound microscope.

MATERIALS

- compound microscope
- prepared slide of mitosis
- paper
- pencil

Procedure

1. View a prepared slide of cells undergoing mitosis under low power of a compound microscope.
2. Move the slide until you find a section where different stages of mitosis are visible.
3. Switch to high power. Use photos or diagrams from your textbook to help you locate and identify cells in interphase and in each stage of mitosis.
4. On a separate piece of paper, sketch an example of each stage. Label each sketch with the following terms where appropriate: *chromosomes, cell membrane, cytoplasm, nucleus, spindle,* and *cell wall.*
5. Switch to low power, and estimate how many cells are clearly in interphase and how many cells are in one of the stages of mitosis.

Analysis

1. **Describe** the activity of chromosomes in each stage of mitosis.

   prophase: chromosomes are visible as dark threads; metaphase: chromatids

   line up along the equator; anaphase: chromosomes appear to pull toward

   opposite poles; telophase: chromosomes are at opposite poles

   _____

   _____

2. **Compare** the number of cells in interphase with the number of cells in one of the stages of mitosis.

   Answers will vary but should indicate more cells in interphase than in the

   other stages.

Name _____ Class _____ Date _____

Observing Mitosis and Cytokinesis *continued*

3. Critical Thinking

Inferring Relationships What does your answer to item 2 indicate about the relative length of interphase?

Cells spend the majority of their time in interphase.

_____

_____

Name _____ Class _____ Date _____

Exploration Lab

Modeling Mitosis

SKILLS

• Modeling

• Using scientific methods

OBJECTIVES

• **Describe** the events that occur in each stage of mitosis.

• **Relate** mitosis to genetic continuity.

MATERIALS

• pipe cleaners of at least two different colors

• yarn

• wooden beads

• white labels

• scissors

Before You Begin

The cell cycle includes all of the phases in the life of a cell. The **cell cycle** is a repeating sequence of cellular growth and division during the life of an organism. Mitosis is one of the phases in the cell cycle. **Mitosis** is the process by which the material in a cell's nucleus is divided during cell reproduction. In this lab, you will build a model that will help you understand the events of mitosis. You can also use the model to demonstrate the effects of **nondisjunction** and **mutations.**

1. Write a definition for each boldface term in the paragraph above and for the following terms: chromatid, centromere, spindle fiber, cytokinesis. Use a separate sheet of paper. **Answers appear in the TE for this lab.**

2. Where in the human body do cells undergo mitosis?

   everywhere in the human body except in most nerve and skeletal muscle cells

3. How does a cell prepare to divide during interphase of the cell cycle?

   During interphase, the cell grows, duplicates its chromosomes and

   organelles, and assembles microtubules.

4. Based on the objectives for this lab, write a question you would like to explore about mitosis.

   Answers will vary. For example: How many chromosomes will each new

   nucleus have after mitosis has occurred?

64

Name _____ Class _____ Date _____

Modeling Mitosis *continued*

Procedure

PART A: DESIGN A MODEL

1. Work with the members of your lab group to design a model of a cell that uses the materials listed for this lab. Be sure your model cell has at least two pairs of chromosomes and is about to undergo mitosis.

> **You Choose**
>
> As you design your model, decide the following:
>
> **a.** what question you will explore
>
> **b.** how to construct a cell membrane
>
> **c.** how to show that your cell is diploid
>
> **d.** how to show the locations of at least two genes on each chromosome
>
> **e.** how to show that chromosomes are duplicated before mitosis begins

2. Write out the plan for building your model. Have your teacher approve the plan before you begin building the model.

The cell model the students build may vary. For example: The students may

use the yarn to represent the cell membrane and spindle fibers, the pipe

cleaners to represent the chromosomes, the beads to represent the cen-

tromeres, and the labels to indicate the genes on the chromosomes.

3. Build the cell model your group designed. **CAUTION: Sharp or pointed objects can cause injury. Handle scissors carefully. Promptly notify your teacher of any injuries.** Use your model to demonstrate the phases of mitosis. Draw and label each phase you model.

4. Use your model to explore one of the questions written for step 4 of **Before You Begin.** Describe the steps you took to explore the question.

Answers will vary.

_____

_____

_____

_____

_____

Name _____ Class _____ Date _____

| Modeling Mitosis *continued*

PART B: TEST HYPOTHESES

Answer each of the following questions by writing a hypothesis. Use your model to test each hypothesis, and describe your results.

5. Cytokinesis follows mitosis. How will the size of each new cell that is formed following cytokinesis compare with that of the original cell?

Each new cell will initially be smaller than the original cell that divides.

6. Sometimes two chromatids fail to separate during mitosis. How might this failure affect the chromosome number of the two new cells?

Nondisjunction of one of the chromosomes will result in one of the new cells

having two copies of that chromosome and the other cell having none.

7. A mutation is a permanent change in a gene or chromosome. What effect might a mutation in a parent cell have on future generations of cells that result from the parent cell?

All of the cells in the subsequent generations of cells will carry the

mutation.

PART C: CLEANUP AND DISPOSAL

8. Dispose of paper and yarn scraps in the designated waste container.

9. Clean up your work area and all lab equipment. Return lab equipment to its proper place. Wash your hands thoroughly before you leave the lab and after you finish all work.

Analyze and Conclude

1. Analyzing Results How do the nuclei you made by modeling mitosis compare with the nucleus of the model cell you started with? Explain your result.

The nuclei students made should be the same as the nucleus of the original

cell except that the chromosomes in the original cell are not replicated until

right before mitosis.

66 Chromosomes and Cell Reproduction

Name _____ Class _____ Date _____

Modeling Mitosis *continued*

2. **Evaluating Methods** How could you modify your model to better illustrate the process of mitosis?

   Answers will vary. Students could mention that the pipe cleaners do not

   show how the shapes of the chromosomes change as they are pulled apart.

   _____

   _____

3. **Recognizing Patterns** How does the genetic makeup of the cells that result from mitosis compare with the genetic makeup of the original cell?

   The genes found in the cells that result from mitosis are the same as the

   genes in the original cell. The chromosomes replicate before cell division.

   One copy of each gene goes to each new cell.

   _____

4. **Inferring Conclusions** How is mitosis important?

   Answers will vary. Students should mention that mitosis preserves the

   chromosome number and genetic makeup of cells.

   _____

5. **Further Inquiry** Write a new question about mitosis or the cell cycle that could be explored with your model.

   Answers will vary. For example: What happens if the DNA is not replicated

   before mitosis begins?

   _____

Exploration Lab

OBSERVATION

Mitosis

Teacher Notes

TIME REQUIRED Part 1: One 45-minute period; Part 2: One 45-minute period

SKILLS ACQUIRED

Identifying and recognizing patterns
Inferring
Interpreting

RATINGS

Easy ◄——1———2———3———4——► Hard

Teacher Prep–2
Student Setup–1
Concept Level–2
Cleanup–2

THE SCIENTIFIC METHOD

Make Observations Students make observations about cells undergoing mitosis.

Analyze the Results Analysis questions 1 and 2 require students to analyze their results.

Draw Conclusions Conclusions question 1 asks student to draw conclusions from their data.

MATERIALS

Materials for this lab can be purchased from WARD'S. See the *Master Materials List* for ordering instructions.

SAFETY CAUTIONS

• Discuss all safety symbols and caution statements with students.

• Instruct students to follow the proper procedures for carrying and using the compound light microscope.

TECHNIQUES TO DEMONSTRATE

Review the use of a microscope, especially the use of the coarse-adjustment and fine-adjustment knobs when viewing under low and high power.

TIPS AND TRICKS

Preparation

If you do not have prepared slides available, you can make your own. Use onion root tips for the plant cell slide. You can obtain animal cells from cheek cells from a fresh fish by using a cotton swab swipe across gill tissue or inside the mouth of the fish.

    a. Rub a small amount of animal tissue from the swab onto a clean slide. For the plant cells, place an onion root tip about 2 mm long on the slide.

    b. Use a wooden macerating stick to break up the cells on the slide.

    c. Place 1–3 drops of water or saline solution over the specimen (if using a stain, apply it now).

    d. Slide the coverslip at a 45° angle along the slide surface. Drop the coverslip onto the tissue once the water or saline has spread across the coverslip.

    e. Use the eraser end of a pencil to press gently on the coverslip to squash the cells on the slide to a one-cell thickness.

Students are sometimes confused while viewing slides of an actual structure when they are used to seeing only an illustration. You may wish to prepare a set of projectable slides showing an illustration of each stage of mitosis side by side with a slide of the actual structures. Allow students to comment and ask questions that will help them orient themselves to the realistic view.

Make students aware of the differences among the stages of mitosis before they begin the lab. A major difficulty in determining a cell's stage of mitosis arises when only part of the cell is visible because of its orientation on the slide, such as when having a polar view from the end of the spindle.

For Extensions item 1, the calculation would be a simple proportion. For example, if 16 out of 200 cells are in metaphase, then 8 percent of the cells are in metaphase. Because 8 percent of 80 minutes is 6.4 minutes, this would be the amount of time that metaphase takes.

Name _____ Class _____ Date _____

Exploration Lab   **OBSERVATION**

Mitosis

All cells undergo a process of growth and division called the *cell cycle*. The cell cycle consists of three major stages: *interphase, mitosis,* and *cytokinesis.* During interphase, the cell grows and the cell's DNA replicates. The next phase is mitosis, during which the replicated genetic material separates into two identical nuclei. Mitosis is divided into four stages: *prophase, metaphase, anaphase,* and *telophase.* Cytokinesis, the last stage of cell division, is the division of the cell cytoplasm between the two new, genetically identical daughter cells.

In this lab, you will study and compare mitosis in animal cells and plant cells. You will also identify cells in the different stages of mitosis.

OBJECTIVES

Observe and **identify** cells undergoing the process of mitosis.

Compare the stages of mitosis in plant cells with these stages in animal cells.

MATERIALS

- compound light microscope
- prepared slides (longitudinal sections of onion or garlic root tips)
- prepared slides (whitefish blastula or other animal tissue)

Procedure

PART 1: MITOSIS IN PLANT CELLS

1. Observe a prepared slide of a longitudinal section of onion or garlic root tips. Use low power to locate the region of actively dividing cells near the end of the root. Using high power, examine individual cells in the region. Find a cell from each stage of mitosis, using the descriptions below. In the space next to each description, draw a cell in that stage as it appears on your slide. Label the visible part of the cell.

Interphase

This is the phase of normal cell activity. During interphase, individual chromosomes cannot be distinguished. Instead they appear as a dark mass of material called *chromatin.* The DNA of each chromosome replicates at the end of this stage. Note the nucleus with one or more dark-stained nucleoli filled with chromatin.

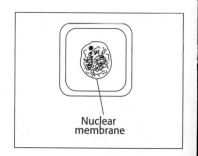

Nuclear membrane

Name _____ Class _____ Date _____

Mitosis *continued*

Prophase

The chromatin appears as a mass of thick threads. These threads are the replicated chromosomes, which have coiled up and shortened. Each chromosome consists of a pair of *chromatids*, which are duplicates of the original chromosome. The chromatids are held together by a centromere. In late prophase, the chromosomes are distinctly visible as pairs of chromatids in the central region of the cell.

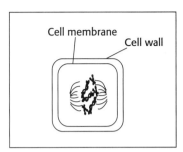

Cell membrane   Cell wall

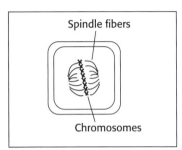

Spindle fibers

Chromosomes

Metaphase

The chromosomes line up across the equator of the cell. A mass of fibers called a *spindle* has formed between the poles of the cell and the mass of chromosomes. A spindle fiber from each pole attaches to each pair of chromatids.

Anaphase

The centromere of each chromatid pair divides. The chromatids move along the spindle fibers toward the poles of the cell. Each chromatid in the pair of chromatids moves toward opposite poles of the cell.

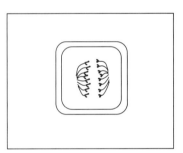

Telophase

The chromatids (now called chromosomes) have formed clumps at each pole. A new nuclear membrane forms around the chromosomes, which uncoil and return to the chromatin network seen in interphase. In plants, the new cell walls grow to form the two new, identical daughter cells.

2. Describe the shape of the cells and the color of the chromosomes.

The cells are rectangular. The chromosomes are stained pink or purple,

depending on the stain used.

Name _____ Class _____ Date _____

Mitosis *continued*

PART 2: MITOSIS IN ANIMAL CELLS

3. Observe the prepared slide of the whitefish blastula or other animal cells. Use low power to locate the cells, then examine them under high power. Find a cell from each stage, using the descriptions below. In the space near each description, draw a cell in that stage as it appears on your slide. Label the visible parts of the cell.

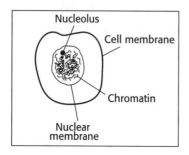

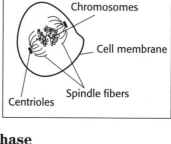

Interphase

A distinct nucleus and a nucleolus are visible. The genetic material appears as chromatin.

Prophase

In early prophase, astral rays have formed around the centrioles, and the spindle has formed between them. The paired chromatids are becoming visible, and the nuclear membrane has disappeared. In late prophase, the chromosomes are short and thick, and are distinct in the central region of the cell.

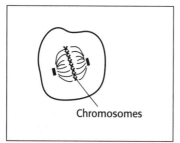

Metaphase

The chromosomes line up at the center of the cell, along the equator.

Anaphase

The chromatids separate at their centromere and are pulled to opposite poles along the spindle fibers.

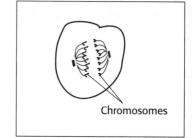

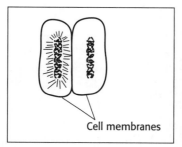

Telophase

The chromosomes appear in clusters at the poles. The parent cell begins to elongate, and the nuclear membranes reform around the chromosome clusters. The spindle and chromosomes become less distinct. The cytoplasm pinches in until the two daughter cells separate during late telophase.

Name _____ Class _____ Date _____

Mitosis *continued*

4. Describe the shape of the chromosomes as they are pulled to the poles in anaphase.

They are V-shaped.

5. Clean up your work area and wash your hands before leaving the lab.

Analysis

1. Identifying Relationships How does mitosis differ in plant cells and animal cells?

During interphase, centrioles appear in the animal cell but not in the plant

cell. During prophase in an animal cell, the astral rays appear around the

centrioles. During telophase in animal cells, the parent cell is pinched in two.

At the end of telophase in plants, a cell wall grows between the two clumps

of chromosomes.

2. Analyzing Data Which phase of mitosis shows the greatest difference between animal cells and plant cells? Explain your choice.

Most students will choose telophase because walls are formed in the plant

cell, which give the cells a rectanglular shape, while the animal cell is

pinched in two, giving the cells an oval shape.

Conclusions

1. Drawing Conclusions What role do you think mitosis plays in living things? Justify your answer.

Students may infer that new, identical cells are being generated and that

these cells are involved in the growth (and renewal) of the organism.

Extensions

1. Building Models Mitosis in onion root tip cells takes about 80 minutes. If you view a slide of a root tip and count the number of cells in each stage of mitosis, you can then calculate the amount of time each stage takes. This is because the percentage of the cells in a particular stage of mitosis is equal to the percentage of 80 minutes that the stage takes. Using this information, devise a method for calculating the amount of time each stage of mitosis takes.

2. Research and Communications Research the length of time mitosis takes in plant cells and in animal cells.

Skills Practice Lab

CONSUMER

Preparing a Root Tip Squash

Teacher Notes

TIME REQUIRED One 45-minute period

SKILLS ACQUIRED

Collecting data
Identifying patterns
Inferring
Interpreting
Organizing and analyzing data

RATINGS

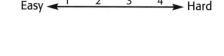

Teacher Prep–3
Student Setup–3
Concept Level–3
Cleanup–3

THE SCIENTIFIC METHOD

Make Observations Procedure steps 13–17 ask students to make observations.

Form a Hypothesis Procedure step 1 asks students to form a hypothesis.

Analyze the Results Analysis questions 1–5 require students to analyze their results.

Draw Conclusions Conclusions questions 3–7 require students to draw conclusions from their data.

MATERIALS

Materials for this lab can be purchased from WARD'S. See the *Master Materials List* for ordering instructions.

SAFETY CAUTIONS

- Discuss all safety symbols and caution statements with students.

- Instruct students to follow the proper procedures for carrying and using the compound microscope.

- In case of contact with aceto-orcein stain, students should flush the affected areas with water for 15 minutes, including under the eyelids, and rinse their mouth with water.

DISPOSAL

- To dispose of volumes of hydrochloric acid and aceto-orcein stain less than 250 mL, do the following. **CAUTION: Wear safety goggles, a face shield, lab apron, and gloves.** Neutralize the acid by adding small amounts of 1 M NaOH solution as required. Place a beaker containing this neutralized solution in the sink, and run water to overflowing for 10 minutes, flushing to a sanitary sewer.

Preparing a Root Tip Squash *continued*

- Prepare and provide separate containers for the disposal of root tips, acid, stain, and broken glass.

TIPS AND TRICKS

This lab works best in groups of two to four students.

If you do not want students handling chemicals, you can stain the root tips yourself. See the last paragraph on this page.

Students will need the entire lab period to complete the procedure. You may want to assign the Analysis and Conclusions questions for homework.

Prior to the start of the lab, review the process of mitosis. Identify the name, events, and chromosome positions of prophase, metaphase, anaphase, and telophase. Have students identify each phase of mitosis in a prepared slide.

Review the parts of the cell. Remind students that nitrogen is part of a plant's DNA and RNA. Phosphate is also part of a cell's DNA and helps plants use water efficiently. Potassium is necessary for photosynthesis and helps make plants hardy.

You can expand the lab by including fertilizer solutions mixed at 50% and 200% concentration in addition to the 100% used in the procedure.

Preparing Fertilizer Solution

Use an all-purpose plant food, such as 15-30-15, that can be found in garden nurseries. (The numbers represent nitrogen, phosphorus, and potassium, respectively.) Prepare one gallon of fertilizer solution, according to the manufacturer's directions, usually 1 tablespoon per gallon of water. The control should contain only water. Label one 500 mL beaker "Y—100%" and one "C—Control."

Fixing and Staining Onion Root Tips

Place five fresh, green onions (scallions) in each of the labeled beakers, and add enough of the appropriate solution to cover the roots. Allow the onions to grow several days until the shortest roots are about 2 cm long.

While the onions are growing, prepare the fixative solution. Mix 25 mL of glacial acetic acid with 75 mL of 95% ethanol.

After the onions grow, cut off about 2 cm of the root tips from each beaker and place the tips in labeled vials containing 100 mL fixative solution. Leave them in a fume hood for 48 hours. **CAUTION: Be sure there are no sources of flame or ignition in the room.** Now, the tips are ready for students to use.

You can store the root tips for up to 2 months by removing them from the fixative solution and placing them in labeled containers with 100 mL of 70% ethanol solution (70 mL of 95% ethanol mixed with 30 mL of distilled water).

To stain the onion tips yourself, add enough 1 M HCl (8 mL of concentrated HCl diluted in 92 mL of distilled water) to each vial to cover the root tips. Let stand for 10 minutes. *Note: If the tips had been preserved in ethanol, place the vials into a 60°C water bath for 15 minutes.* Use a bulb pipet to draw off as much HCl from the vials as possible. Add enough aceto-orcein stain to the vials to cover the root tips completely. After about 15 minutes, the root tips will turn bright pink and are ready for students to use to prepare a squash. Therefore, in step 3, students can place a tip directly onto a microscope slide on a paper towel. They can proceed to steps 8, 10, and the rest of the lab.

Name _____ Class _____ Date _____

CONSUMER

Preparing a Root Tip Squash

To observe mitosis in stem and root meristems, biologists prepare a special slide called a *squash.* This preparation is just what it sounds like. Actively dividing cells from a root or stem meristem are removed and treated with hydrochloric acid to fix the cells or to stop them from dividing. The cells are then stained, made into a wet mount, and squashed. Squashing spreads the cells into a single layer.

In this lab, you will study mitotic activity in the roots of onion plants to determine the effects of fertilizer on actively dividing root cells. You also will determine the percentage of cells that are undergoing mitosis. To observe mitosis in the onion root cells, you will make a squash of onion root tips, which contain actively dividing cells.

OBJECTIVES

Prepare and stain slides of onion root tips.

Observe cells in the process of mitosis.

Form and **test** a hypothesis about the effects of fertilizer on mitosis.

Determine the effect of fertilizer on the percentage of cells in mitosis.

MATERIALS

- aceto-orcein stain in dropping bottle (30 mL)
- coverslips (4)
- compound light microscope
- distilled water
- eyedropper
- forceps
- gloves
- 1 M HCl in dropping bottle (6 mL)

- lab apron
- microscope slides (4)
- onion root tips, specially grown
- paper (2 sheets)
- paper towels
- petri dish
- safety goggles
- wooden macerating stick

Procedure

1. Write a hypothesis that clearly states how you think the concentration of fertilizer affects mitosis in onion roots.

   Hypotheses will probably predict that the presence of fertilizer will result in

   a greater percentage of mitotic activity than without fertilizer.

PREPARING THE SQUASHES

2. Put on safety goggles, gloves, and a lab apron.

3. Use forceps to carefully remove three onion root tips from the vial labeled "Y—100%," and place them in a petri dish.

4. Use an eyedropper to flood the root tips with 1 M HCl. Allow the root tips to stand in the HCl for 10 minutes. **CAUTION: HCl is an irritant and a poison. Avoid skin and eye contact. If you get HCl on your skin or clothing, wash it off at the sink while notifying your teacher. If you get HCl in your eyes, promptly flush it out at the eyewash station while notifying your teacher. Notify your teacher in the event of an acid spill.**

5. Use the eyedropper to remove the HCl from the petri dish, and dispose of it as instructed by your teacher. Be careful to not remove the root tips with the eyedropper. Refill the petri dish with distilled water.

6. Place a microscope slide on a paper towel. Add three drops of aceto-orcein stain to the center of the slide. **CAUTION: Aceto-orcein stain is corrosive and poisonous. Avoid eye and skin contact. In case of contact, notify your teacher immediately. Avoid inhaling the vapors. Aceto-orcein stain will also stain your skin and clothing. Promptly wash off spills to minimize staining.**

7. Use forceps to transfer a prepared root tip from the petri dish to the drop of stain on the microscope slide.

8. Pulverize the tissue by gently but firmly tapping the root tip with the end of a wooden macerating stick. *Note: Move the stick in a straight up-and-down motion.*

9. Allow the root tip to stain for 10–15 minutes. *Note: Do not let the stain dry. Add more stain if necessary.*

10. Place the slide on a smooth, flat surface. Add a coverslip to the slide to make a wet mount. Place the wet mount between two pieces of paper towel.

11. Use the eraser end of a pencil to press down on the coverslip. Apply only enough pressure to squash the root tip into a single cell layer. Be very careful not to move the coverslip while you are pressing down with the pencil. Do not press too hard because you might break the glass slide or tear apart the cells.

12. Repeat steps 3 through 11 for the C—Control root tips. Place each squash on a piece of paper labeled "C—Control."

Name _____ Class _____ Date _____

Preparing a Root Tip Squash *continued*

OBSERVING THE SQUASHES

13. View one slide at a time with a compound light microscope under both low and high power. *Note: Remember that your mount is fairly thick, so be careful not to change to the high-power objective too quickly. Doing so could shatter the coverslip and destroy your preparation. You will need to focus carefully with the fine-adjustment knob to see the structures under study.* View all three slides you have made. Select the slide that shows the most cells undergoing mitosis. Use this slide to complete steps 14 and 15.
   • What is the shape of the onion root tip cells?

   Most cells will appear rectangular. Some cells will be almost square in shape.

   • What color did the aceto-orcein make the chromosomes?

   The stain was dark red.

14. Observe the slide under high power. Without moving the slide, estimate the number of cells in the viewing area. To estimate, mentally divide the viewing area into three viewing sections. Then count the cells in one section and record the number for each viewing section in **Table 1.** Total the three numbers and record in the table.

15. In each viewing section, count the number of cells in prophase, metaphase, anaphase, and telophase. Record the numbers in **Table 1.**

TABLE 1 STAGES OF MITOSIS

	Viewing section							
	1		**2**		**3**			
	Y	**C**	**Y**	**C**	**Y**	**C**	**Total Y**	**Total C**
Total number of cells		285		285		285		855
							Total number of cells per stage	
Prophase	19	6	15	4	22	8	56	18
Metaphase	8	2	16	4	12	3	36	9
Anaphase	12	3	9	2	8	2	29	7
Telophase	6	1	5	2	8	1	19	4

Entries will vary for each group. Sample data are entered above.

Name _____ Class _____ Date _____

Preparing a Root Tip Squash *continued*

- What are the most common stages of mitosis you observed?

  Most students will see prophase and anaphase, but all stages of mitosis

  will be present.

16. For the next observation, move the slide to a new viewing section next to the one you just looked at. Repeat steps 14 and 15 in this new section. Repeat steps 14 and 15 for the third viewing section.

17. Repeat steps 14–16 for the C-Control slide you have prepared.

18. Dispose of your materials according to your teacher's instructions. Clean up your work area, and wash your hands before leaving the lab.

Analysis

1. Explaining Events Why do specimens have to be thin to be viewed through the microscope?

Internal structures cannot be identified if the specimen is more than one cell

layer thick. Also light must be able to pass through the specimen.

2. Describing Events How many cells were in mitosis in each of the slides?

Answers will vary. Sample data include 140 cells grown in fertilizer solution

and 38 cells grown in the control solution.

3. Analyzing Data What percentage of cells on your slide were in the process of mitosis? Use the following formula to calculate your answer. Show your calculations in the space provided.

$$\% \text{ cells in mitosis} = \frac{\text{total number of cells in all phases of mitosis}}{\text{total number of cells}} \times 100$$

Answers will vary, but the control sample should resemble the following: number of cells in mitosis = 38, total number of cells = 855, % of cells in mitosis = 38/855 × 100 = 4.4%.

4. Analyzing Data What percentage of cells on each slide were in each phase of mitosis? Use the following formula to calculate your answer.

$$\% \text{ cells in phase} = \frac{\text{total number of cells in phase}}{\text{total number of cells}} \times 100$$

Answers will vary, but the control sample should approximate the following: prophase = 18/855 = 2%, metaphase = 9/855 = 1.1%, anaphase = 7/855 = 0.8%, telophase = 4/855 = 0.5%.

Name _____ Class _____ Date _____

Preparing a Root Tip Squash *continued*

5. **Explaining Events** Why did you make two slides in this lab?

One slide was prepared from root tips grown in fertilizer. One slide was a

control grown in only water. Preparing two slides allows a comparison of the

percentage of cells in mitosis and the effects of fertilizer on mitosis.

Conclusions

1. **Evaluating Methods** Why do you squash and spread out the root tip?

Squashing makes the tissue one cell layer thick.

2. **Interpreting Information** How do you explain the low percentage of cells undergoing mitosis in each slide?

The number of cells undergoing mitosis seems low because most of the cell's

life is spent at rest or in interphase.

3. **Drawing Conclusions** Which slide showed the greatest number of cells in mitosis?

The slide of root tips grown in fertilizer is most likely to result in the great-

est number of cells in mitosis.

4. **Evaluating Methods** Why is a root tip a good choice for studying the effects of fertilizer on mitosis?

The root tip is one of the areas that is growing most rapidly. The cells will be

undergoing mitosis at a higher rate in this area than in other regions of the

plant. Because the root tip contains no chlorophyll, cells undergoing mitosis

are easy to see.

Name _____ Class _____ Date _____

Preparing a Root Tip Squash *continued*

5. **Defending Conclusions** What can you conclude about the effect of fertilizer on the growth of an onion plant?

Answers will vary, but should include that fertilizer increases growth rate as

shown in this experiment.

6. **Making Predictions** Consider the statement "If some is good, more is better." Predict a possible result if you were to double the concentration of fertilizer solution.

Answers will vary. Students may predict that a high concentration of fertilizer

could be harmful to a plant. Students may also predict that more fertilizer

will make the plant grow faster.

7. **Evaluating Results** Was your hypothesis in step 1 supported? Explain.

Answers will vary, according to each student's hypothesis. Some students will

have formed a hypothesis that fertilizer will stimulate a greater amount of

mitosis in the root tip.

Extensions

1. **Designing Experiments** Design an experiment to test the effect of different formulas of fertilizer on the percentage of cells undergoing mitosis in an onion root tip. For example, identical concentrations of fertilizer could be tested by using different formulas of fertilizer such as 15-30-15, 10-16-16, 30-10-10, or nitrogen only. Be sure to control all variables other than the fertilizer.

2. **Research and Communications** Botanists are biologists who study plants. They study plant structure, function, growth, and reproduction. Because there are so many things to know about so many different kinds of plants, many botanists specialize in studying just one function of just one kind of plant. Find out about the training and skills required to become a botanist.

Answer Key

Directed Reading

SECTION: CHROMOSOMES

1. Cell division is the process by which new cells form. Gametes are reproductive cells that form by one type of cell division.
2. A gene is a segment of DNA that codes for a protein or RNA molecule. DNA is a substance made of nucleotides that stores information about when and how to make proteins.
3. Chromosomes are structures made of DNA and associated proteins. Chromatids are the two parts of a chromosome that contain identical copies of DNA and are joined by a centromere.
4. 4
5. 2
6. 5
7. 3
8. 1
9. homologous
10. diploid
11. n
12. zygote
13. Autosomes are chromosomes that are not directly involved in determining the sex of an individual. Sex chromosomes, such as X and Y chromosomes, contain genes that determine the sex of an individual.
14. A karyotype is a photograph of the chromosomes in a dividing cell that shows the chromosomes arranged by size.
15. deletion—a piece of a chromosome breaks off; duplication—a chromosome fragment attaches to its homologous chromosome, which then has two copies of the genes in the fragment; inversion—a chromosome fragment attaches to the same chromosome in the reverse orientation; translocation—a chromosome fragment attaches to a nonhomologous chromosome

SECTION: THE CELL CYCLE

1. division
2. interphase
3. e
4. d
5. c
6. a
7. b
8. proteins
9. cell growth
10. genes
11. Mutations in genes that produce checkpoint proteins may cause the proteins to malfunction and prevent the checkpoints from controlling cell division. Uncontrolled cell division may result in cancer.

SECTION: MITOSIS AND CYTOKINESIS

1. They move chromosomes during mitosis.
2. They help form the spindle.
3. c
4. b
5. a
6. d
7. 1
8. 4
9. 2
10. 3
11. after
12. a belt of proteins

Active Reading

SECTION: CHROMOSOMES

1. A gene is a segment of DNA that codes for a protein or RNA molecule.
2. The strand is stretched out so that the information it contains can be decoded and used to direct the synthesis of proteins needed by the cell.
3. Chromatids are exact copies of DNA that make up chromosomes.
4. b

SECTION: THE CELL CYCLE

1. cell cycle, interphase
2. cell cycle: repeating sequence of growth and division during the life of a cell; interphase: first three phases of the cell cycle
3. It is in the process of dividing.
4. d
5. a
6. e
7. c
8. a
9. d
10. b
11. c
12. b
13. a
14. d
15. Both are phases of the cell cycle in which a cell part divides. However, during mitosis, a nucleus divides, while during cytokinesis, cytoplasm divides.
16. d

SECTION: MITOSIS AND CYTOKINESIS

1. p	5. b
2. b	6. p
3. a	7. d
4. p	

Vocabulary Review

1. c	11. b
2. a	12. c
3. b	13. b
4. d	14. a
5. c	15. b
6. d	16. c
7. a	17. a
8. c	18. c
9. a	19. d
10. b	

Science Skills

SEQUENCING

1. b
2. e
3. g
4. a
5. d
6. h
7. c

8. f
9. anaphase
10. telophase
11. metaphase
12. prophase
13. 3
14. 2
15. 4
16. 1

Concept Mapping

1. prokaryotes
2. eukaryotes
3. growth
4. repair
5. asexual reproduction
6. sexual reproduction
7. first growth phase
8. synthesis phase
9. second growth phase
10. cytokinesis
(Items 3–6 are interchangeable. Items 7–10 are interchangeable.)

Critical Thinking

1. a	14. b
2. d	15. a
3. e	16. f
4. b	17. b, f
5. c	18. i, d
6. e	19. j, h
7. a	20. g, a
8. c	21. c, e
9. d	22. c
10. b	23. c
11. d	24. b
12. e	25. a
13. c	

Test Prep Pretest

1. c
2. b
3. b
4. d
5. b
6. a
7. b
8. gene
9. Down syndrome
10. autosomes
11. sex chromosomes

12. cell cycle
13. DNA synthesis
14. interphase
15. microtubule
16. cancer
17. copied
18. Prior to cell division, the DNA and proteins associated with the DNA coil into structures called chromosomes.
19. A frog somatic cell is diploid and therefore has twice the number of chromosomes of the haploid frog egg cell.
20. When one or more chromosomes fail to separate properly during cell division in gamete production, one new cell will receive both chromosomes, and the other new cell will receive none. As a result, one gamete will be produced that has one extra chromosome, and another gamete will be produced that is missing one chromosome.
21. If conditions are favorable for cell division during the first growth phase, certain proteins will stimulate the cell to begin the synthesis phase, during which DNA is duplicated. During the second growth phase, the DNA replication is checked by DNA repair enzymes. If everything is in order, proteins then initiate the beginning of mitosis. At the final checkpoint, the cell is prompted to exit from the mitosis phase and to begin the first growth phase again.
22. prophase, metaphase, anaphase and telophase
23. During prophase, chromosomes coil up and become visible, the nuclear envelope dissolves, and the spindle apparatus forms. During metaphase, the chromosomes line up in the center of the cell. During anaphase, the centromeres divided and the chromatids are drawn to opposite poles by spindle fibers. During telophase, a nuclear envelope forms around the chromosomes at each pole. The chromosomes uncoil, and the spindle dissolves.

24. The phases of interphase are the first growth (G_1) phase, the synthesis (S) phase, and the second growth (G_2) phase. The G_1 phase is the growth phase of a cell when the cell grows rapidly and carries out its routine functions. DNA is copied during the S phase. In the G_1 phase, preparations are made for nuclear division. Proteins needed for cell division are synthesized.
25. growth, development, repair, reproduction

Quiz

SECTION: CHROMOSOMES

1. d	6. c
2. b	7. b
3. e	8. c
4. c	9. b
5. a	10. d

SECTION: THE CELL CYCLE

1. d	6. b
2. e	7. a
3. b	8. c
4. a	9. b
5. c	10. b

SECTION: MITOSIS AND CYTOKINESIS

1. e	6. c
2. a	7. b
3. d	8. d
4. c	9. a
5. b	10. d

Chapter Test (General)

1. f	11. d
2. e	12. b
3. h	13. c
4. i	14. c
5. a	15. b
6. c	16. d
7. d	17. a
8. g	18. d
9. j	19. a
10. b	20. d

Chapter Test (Advanced)

1. c
2. a
3. e
4. d
5. b
6. c
7. c
8. d
9. c
10. a
11. zygote
12. binary fission
13. interphase
14. gametes
15. spindle
16. genes, proteins
17. homologous chromosomes
18. genes, chromosomes
19. deletion mutation
20. DNA
21. During anaphase, the spindle helps to separate the chromatids by dragging them to the opposite poles of the cell.
22. A karyotype is a photograph that shows the collection of chromosomes found in an individual's cells. Analysis of this collection of chromosomes can reveal abnormalities in chromosome number and structure. Down syndrome is associated with trisomy 21, an extra chromosome 21 in a person's cells.
23. The G_1 state of the cell cycle is the phase of cell growth. This is followed by the S stage, in which DNA is copied. G_2 involves the cell preparing for cell division. During mitosis, the nucleus of a cell is divided into two nuclei with each nucleus containing the same number and kinds of chromosomes as the original cell. The cell cycle continues with cytokinesis, the stage during which the cytoplasm divides.
24. Animal cells lack cell walls. In animal cells, the cytoplasm is divided when a belt of protein threads pinches the cell in half. In plant cells, the Golgi apparatus forms vesicles that fuse in a line along the center of the cell and form a cell plate. A new cell wall then forms on each side of the cell plate.
25. If proteins that normally slow or stop the cell cycle are inactivated, the cell cycle would continue indefinitely, resulting in uncontrolled cell division. This uncontrolled division of cells could result in cancer.